FIVE EVIDENCES FOR A

GLOBAL FLOOD

BRAD FORLOW

FIVE EVIDENCES FOR A

GLOBAL FLOOD

INSTITUTE
for CREATION
RESEARCH

Dallas, Texas
www.icr.org

FIVE EVIDENCES FOR A GLOBAL FLOOD
by Brad Forlow, Ph.D.

All Scripture quotations are from the King James Version.

ISBN: 978-1-935587-07-1

Please visit our website for other books and resources: www.icr.org

Printed in the United States of America.

TABLE OF CONTENTS

INTRODUCTION

Water is a powerful force. In recent years, earthquakes have caused upheavals under the ocean floor and produced massive tsunamis that have killed and displaced tens of thousands of families. Even torrential rainfall in the lowlands have swelled rivers beyond their banks and flooded hundreds of miles of farmland, homes, and downtown urban areas. And, of course, hurricanes have hit the coasts with such force that the rain, waves, and wind have devastated local populations so badly that it requires years for recovery.

But what of the idea of a flood of water that could devastate the entire earth, even covering the tops of the highest mountains? Is such a flood possible? And if so, what forces would be necessary to make this a reality? Is there proof?

While a worldwide flood seems unimaginable to us today, there is evidence that such a flood did occur thousands of years ago—a catastrophe of such magnitude that almost every creature with breath in its nostrils was killed.

The idea of a flood so powerful that it could cover every square inch of the earth is unthinkable today. But the fact is that there is solid evidence to verify that this flood did occur—evidence from Scripture as well as evidence from science.

There is evidence that demonstrates the details of and purpose for what is known to us as Noah's Flood. There is evidence that is drawn from historical accounts not only in Genesis, but throughout the entire Bible. And there is solid evidence that is seen through studies in geology and paleontology.

So, the question remains: Why would a flood that ravaged the earth thousands of years ago be relevant to people today?

Some of the answers for this question have run the extremes of human thought.

- "I think the story of Noah and the Ark is just a myth. There are a lot of those in the Bible."

- "Every culture has some sort of 'flood' legend. Genesis is just one more story. Good lessons, but just a story."

- "Genesis 1–11 cannot be taken literally; scientific discoveries won't allow it."

- "It really doesn't matter whether the Flood was real or not, or local or global. The important thing is the lesson about salvation."

- "The earth is billions of years old, the geologic column is a scientific fact. So, the Bible just doesn't make the case for such a catastrophe."

The sad reality is that many Christians believe such statements about the Bible, about the Genesis record, and about the Flood. Many who claim to believe in the Bible actually disbelieve much of Genesis, especially about the creation, the Fall of man, and the Flood of Noah's day. If Genesis is mythology, as some Christians teach, then what of the virgin birth or the resurrection?

If the Flood were just a local or regional event, then why the need for the Ark, and why such a

There is much evidence—both biblical and scientific—that must be careful considered by those unafraid to follow the evidence where it leads.

big boat? Noah, and all the people of the region, could simply have moved to higher ground—they certainly had plenty of forewarning. And if the Flood was not global, then why do scientists today find evidence across the globe that suggests a worldwide water catastrophe? How do we account for the topsy-turvy arrangement of fossils throughout the earth's strata? There is much evidence—both biblical and scientific—that must be careful considered by those unafraid to follow the evidence where it leads.

Is the Bible the divinely inspired and inerrant book authored by God, or is it simply a collection of stories by well-meaning, but unlearned, men who just didn't understand all the facts?

Of course, those who fully trust the Word of God know better. They look at the evidence and see an all-powerful Creator who is also Judge and Savior. Most of all, those who see the Bible as absolutely true recognize that God's Word is supremely authoritative—even more so than those guys in white lab coats.

The writer of Hebrews recognized Noah as an example of faith—one who listened to God and obeyed:

> By faith Noah, being warned of God of things not yet seen, moved with fear, prepared an ark for the saving of his house; by which he condemned the world, and became heir of the righteousness which is by faith. (Hebrews 11:7)

There's probably a lot that Noah did not understand, but one thing he knew—he could trust every word God said, and he humbly obeyed his Lord. No doubt Noah and his family were fearfully amazed at the devastation of the torrential rains, the waves, the bursting forth of the fountains of the deep, and the forces that consumed every human being on the earth except themselves. What a memory of divine power and judgment.

But Noah experienced God's grace as well—grace that brought a new beginning on the mountains of Ararat and, in turn, new families and clans and cities and civilizations.

God destroyed life on earth because of sin, but He granted life anew on the earth because of His grace, mercy, and love.

This book is designed to give you five compelling evidences that show the Flood of Noah's day to be absolutely true. Many larger volumes have been written about all these evidences, but this book will give you a handy guide that's easy to digest as you study the Bible and consider the world that God made—and judged.

By the end of the book, you should be able to remember and use these five evidences when you meet those who snicker at the mention of Noah and the Ark.

Evidence 1	Genesis Teaches a Global Flood.
Evidence 2	God Judged the World Through a Global Flood.
Evidence 3	All Scripture Testifies of a Global Flood.
Evidence 4	Geology Reveals the Fact of a Global Flood.
Evidence 5	The Fossil Record Requires a Global Flood.

EVIDENCE 1
GENESIS TEACHES
A GLOBAL FLOOD

The biblical story of Noah's Flood is probably familiar to most people today. But is it just a story, a myth, or a legend? Is it acceptable to teach that the Flood of Noah's day was local because a global flood is not realistic and seemingly impossible? What insights can be gained from a study of Scripture as to the extent of the Flood—local or global? Based upon what is recorded in Genesis, it is evident that the Flood of Noah's day was worldwide. Therefore, those who propose a local flood must "reinterpret" Genesis to dismiss the clear teaching of the Flood's global nature. In Genesis alone, there is vast testimony to a global flood.

The Cause of the Flood

Many wonder if it is even possible for the whole earth to be flooded. Those skeptical might ask, "Where did all the water come from?" That is a common question (or objection), but one that is answered in Genesis. Genesis tells us that the flood waters came from three sources (Genesis 7:11-12; 8:2).

1. The flood waters came from "all the fountains of the great deep" breaking open, referring to water reservoirs under the ocean floor. This would have led to enormous geological

activity, including volcanoes, earthquakes, and tsunamis.

2. The flood waters came from the "windows of heaven" opening, likely referring to a water reservoir in the atmosphere.

3. Rain also fell upon the earth for 40 days and 40 nights. These three sources are all global sources of water.

Based on the text of Scripture, the Flood of Noah's day could hardly be a local flood because global sources of water and global causes of the Flood necessitate a global effect.

The Depth of the Flood

Something else in Genesis that speaks to the extent of the Flood is how high the flood waters rose. Genesis 7:19-20 says, "And the waters prevailed exceedingly upon the earth; and all the high hills, that were under the whole heaven, were covered. Fifteen cubits upward did the waters prevail; and the mountains were covered." In these verses we find that the flood waters rose to a depth of 15 cubits (22.5 feet) above the highest mountains. Many devastating local floods probably do not often exceed water levels of 22.5 feet. But the Flood described in Genesis resulted in waters that rose 22.5 feet above the highest mountains, not just above ground level.

Today, the mountains of Ararat (Genesis 8:4) include Mount Ararat, which rises to 17,000 feet in elevation. Is it reasonable to believe that only a local flood covered this 17,000-foot mountain? The waters of a local flood could not cover all of the mountains with 22.5 feet of water. A 17,000-foot flood is not a local flood!

The Duration of the Flood

Contrary to the popularization of the story of Noah's Flood, the Flood did not last for 40 days and 40 nights. So, how long did it last? We are actually given very precise information regarding its duration. The exact dates (days) and timeline are revealed in the Genesis account. This attention to detail emphasizes the historical nature of the event.

Genesis gives us a clear starting date of the Flood: "In the six hundredth year of Noah's life, in the second month, the seventeenth day of the month" (Genesis 7:11). On that day, the Lord shut Noah and his family in the Ark (Genesis 7:16).

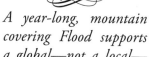

A year-long, mountain covering Flood supports a global—not a local—Flood.

Although it rained for 40 days and 40 nights, the waters continued to rise on the earth for 150 days (Genesis 7:24). The water sources were then restrained and the earth dried out over a period of another 221 days. The exact date of Noah's exit from the Ark is also given. He did not leave the Ark until the six hundred and first year, in the second month, on the twenty-seventh day of the month (Genesis 8:13-14).

The attention to detail in Genesis is astounding. Following the timeline given in the Bible, Noah and his family spent 371 days in the Ark, which means that the Flood lasted over one year! Even the waters from the most devastating local floods of our time recede within a matter of a few weeks. A year-long, mountain-covering flood supports a global—not a local—flood.

The Need for the Ark

Does the fact that God instructed Noah to build an Ark speak to the extent of the Flood? Well, how big was this Ark?

According to Genesis, the Ark built by Noah was approximately 450 feet long, 75 feet wide, and 45 feet high (Genesis 6:15). This results in an inside capacity of around 1.5 million cubic feet. Remarkably, it has been shown scientifically that it would have been practically impossible to capsize the Ark. Isn't this a rather large vessel for eight people and some local animals, if it was only going to be a local flood?

Actually, if it was only going to flood locally, was the Ark even necessary? Couldn't Noah and his family just have walked to a different region, and couldn't the animal have just migrated away? After all, Noah had upwards of 100 years' warning. An Ark large enough to house two or more of each land-dwelling, air-breathing animal testifies to a global flood.

Genesis Explicitly Describes a Global Flood

When reading the Genesis account of the Flood, the extensive explicit "universal" language cannot be missed. Over 30 references in Genesis 6–9 express the global nature of the Flood. Consider the following phrases used in the description of the Flood:

- "The earth also was corrupt…for all flesh had corrupted his way" (6:11-12).
- "The end of all flesh is come….I will destroy them with the earth" (6:13).
- "Destroy all flesh…under heaven" (6:17).
- "Everything that is in the earth shall die" (6:17).
- "Every living substance that I have made will I destroy" (7:4).
- "All the fountains of the great deep" (7:11).
- "The windows of heaven" (7:11).
- "All the high hills, that were under the whole heaven, were covered" (7:19).
- "The waters prevail; and the mountains were covered" (7:20).
- "All flesh died that moved upon the earth…and every man" (7:21).
- "All that was in the dry land, died" (7:22).
- "Every living substance was destroyed which was upon the face of the ground" (7:23).
- "Noah only remained alive, and they that were with him in the ark" (7:23).
- "Neither will I again smite any more every thing living" (8:21).
- "Neither shall all flesh be cut off any more" (9:11).
- "Every living creature of all flesh" (9:15).

The global nature of the language used to describe the extent of the Flood is also validated later in the book of Genesis. A good example of the use of global language is found in Genesis 8:9, which states the waters were on "the face of the whole earth." The same phrase is also used by the author of Genesis in the account of the Tower of Babel in reference to the universal/global dispersion of the people. If the Flood were only a local event, why would the author of Genesis use such explicit "global" language to describe the events and nature of the Flood?

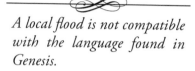

A local flood is not compatible with the language found in Genesis.

A local flood is not compatible with the language found in Genesis. The language chosen to describe the Flood in Genesis clearly communicates the Flood's global nature. If God was really trying to describe a local flood, He obscured the facts, for over and over again the wording demands a global flood.

Genesis presents a "precise set of words and a very specific grammatical construction that clearly identifies the reason for the Flood, the extent of the Flood, and its uniqueness....If God merely intended to destroy the settlements in the Mesopotamian river valley [local flood], He was using grossly exaggerated language. If these words do not mean what they clearly say, then how can we trust any portion of Scripture?"[1] The Bible can be trusted as authoritative truth. The explicit teaching related to the physical causes of the Flood, the depth of the Flood, the duration of the Flood, and the global language used throughout gives strong biblical evidence that the Flood of Noah's day was global.

---

1. Morris III, Henry. 2003. *After Eden*. Green Forest, AR: Master Books, 161.

EVIDENCE 2
GOD JUDGED THE WORLD THROUGH A GLOBAL FLOOD

T he language used to describe the flood in Genesis strongly speaks to a worldwide event. But why did God send the Flood, and does the reason give further evidence of its extent? Not only does Genesis explicitly teach about a flood with global impact, but Genesis also shows that the purpose of the Flood was *universal* judgment.

The Reason for the Flood Was Judgment of All Mankind

God's perfect creation (Genesis 1–2) was corrupted by the sin of Adam and Eve (Genesis 3). Their act of rebellion brought wickedness, evil, and death into the world. In just the first few chapters of Genesis, evil and wickedness increased in the world. Known for murder, Cain and Lamech (Genesis 4) are popular examples of the growing disregard for human life. The marriage between the sons of God and the daughters of men epitomized man's rebellious nature (Genesis 6:1-4). Following this account, the author of Genesis revealed the reason for the Flood:

> And God saw that the wickedness of man was great in the earth, and that every imagination of the thoughts of his

heart was only evil continually. And it repented the LORD that he had made man on the earth, and it grieved him at his heart. And the LORD said, I will destroy man whom I have created from the face of the earth; both man, and beast, and the creeping thing, and the fowls of the air; for it repenteth me that I have made them. (Genesis 6:5-7)

These verses answer the question of why God sent the Flood. He was grieved at the wickedness and evil of man, man's rebellion, and the violence in the world. God sent the Flood as an act of judgment. Did you know that portions of four chapters and over 1,700 words are devoted to the telling of the Flood of Noah's day? Seven different times the "sentence" of judgment is pronounced or referred to in these chapters.

But what is said about the extent of the judgment? In the verses below, note the global or universal language that is used in reference to the judgment being brought upon the earth.

God sent the Flood as an act of judgment.

> The *earth* also was corrupt before God, and the *earth* was filled with violence. And God looked upon the *earth*, and, behold, it was corrupt; for *all* flesh had corrupted his way upon the *earth*. And God said unto Noah, The end of *all* flesh is come before me; for the *earth* is filled with violence through them; and, behold, I will destroy them with the *earth*....And, behold, I, even I, do bring a flood of waters upon the *earth*, to destroy *all* flesh, wherein is the breath of life, from *under heaven*; and *every* thing that is in the earth shall die....For yet seven days, and I will cause it to rain upon the earth forty days and forty nights; and *every* living substance that I have made will I destroy from off the *face of the earth*. (Genesis 6:11-13, 17; 7:4, emphasis added)

Repeatedly, the *earth* is being described as corrupt and filled with violence. Note the language used to describe what would be affected by the judgment—*all, every, under*

The language used to describe the judgment from God is unmistakably global in nature.

heaven, and the *face of the earth*. It is also interesting to note that in Genesis 6:17, the Hebrew word used for flood is *mabbul*. The use of this particular word emphasizes the unique nature of the event. Various other words are used in Scripture for local floods, but the use of *mabbul* speaks to the Flood of Noah's day as being *the* Flood.

The language used to describe the Flood (Evidence 1) indicates a flood of worldwide significance. The reason for the Flood was judgment. Fittingly, the language used to describe the judgment from God is unmistakably global in nature. Therefore, since the wickedness of man brought God to destroy man from the face of the earth, isn't it perfectly reasonable that the response would be a judgment that affected the whole earth and all of mankind—a worldwide Flood?

The Purpose of the Flood Was the Destruction of All Life

The reason for the Flood was universal judgment. Man was wicked and evil, and the earth was characterized by corruptness and violence. But the judgment for the wickedness of man that grieved God also impacted the entire earth. God clearly pronounces destruction, not only of man, but of all the animals and birds on the earth (Genesis 6:5-7). The reason for the Flood was judgment, but the purpose of the Flood was to destroy all life on the earth *and* the earth itself.

> And all flesh died that moved upon the earth, both of fowl, and of cattle, and of beast, and of every creeping thing that creepeth upon the earth, and every man: All in whose nostrils was the breath of life, of all that was in the dry land, died. And every living substance was destroyed which was upon the face of the ground, both man, and cattle, and the creeping things, and the fowl of the heaven; and they were destroyed from the earth: and Noah

only remained alive, and they that were with him in the ark. (Genesis 7:21-23)

Can a flood sent for this reason and to accomplish this purpose be simply a tranquil, local flood? The emphasis in these verses is complete and absolute destruction of all flesh (animals and man) upon the earth. Doesn't this demand a global flood, since men and animals lived beyond the local region of Noah? To accomplish this purpose, a global and catastrophic Flood is necessary.

Read these chapters in Genesis carefully. Does the account of the Flood of Noah's day leave any room for the interpretation of a local flood? The Flood that is described accomplished the complete destruction of all flesh having the breath of life. All animal life on earth was destroyed by the Flood. No one on earth survived the Flood except for Noah and his family. Everyone else died. The Bible does not leave that open to any other interpretation (Genesis 7:23).

> *The world that existed before the Flood was destroyed.*

Not only was all life destroyed from the earth, but the earth itself was forever changed by the Flood. Within the stated purpose of the Flood—to completely destroy all flesh having the breath of life (man and animals)—don't miss God's declaration that it was to destroy the earth as well. God vowed to "destroy them with the earth" (Genesis 6:13). The promise from God was not to destroy "all flesh" *from* the earth, but to destroy them *with* the earth. In the judgment for man's wickedness, God includes the earth over which man was given dominion. Therefore, the world that existed before the Flood was destroyed. The global destruction of the earth by the Flood is validated in the New Testament by Peter (2 Peter 3:6).

God's purpose in the judgment of the Flood was to destroy all air-breathing life on the face of the earth (Genesis 6:13, 17). To destroy "all flesh" along with "the earth" requires a worldwide flood. Both the reason for and the purpose of the Flood give evidence of its global nature.

God's Promise Confirms the Global Flood

When you see a rainbow, are you reminded of the story of Noah's Ark? Do you promptly ask your children what the meaning of the rainbow is? Many people immediately associate the story of Noah's Ark with the rainbow. And that is the point. The rainbow is the sign of the covenant that God made with Noah. It is a covenant that extends to us today (Genesis 9:12-13). What is this covenant, and how does it give further confirmation of a worldwide flood?

When Noah came off the Ark, he offered sacrifices to God. God made a covenant with him, promising to never again send a flood to destroy the earth and all the living creatures (Genesis 8:2; 9:11). This everlasting covenant was made "between God and *every* living creature of *all* flesh that is *upon the earth*" (Genesis 9:16, emphasis added). Once again, the language reiterates the universality of the Flood, because God promises that all flesh and the earth will never again be destroyed by flood waters.

But, consider the argument for a local flood. Since the time of Noah, many local and regional floods have occurred that have caused great destruction and great loss of life. If the Flood of Noah's day was only a local or regional flood, then God has not kept His promise, as there have been numerous devastating local floods, even recently, that have caused drastic loss of life and destruction of the earth. The promise that God made to never flood the earth again validates that the Genesis Flood was worldwide, not local.

Noah's Ark Is a Picture of Salvation

The Bible teaches that the reason for the Flood was universal judgment and that the Flood brought complete destruction of all life on earth and of the earth itself. Therefore, for Noah, his wife, his sons, and his son's wives, the Ark was the means of salvation from the judgment (Genesis 6:8). God gave Noah very specific instructions on

The Ark becomes a beautiful picture of our salvation through Christ's atonement.

how to build the Ark. These instructions included that it was to be covered inside and out with "pitch" (Genesis 6:14). This "covering" made the Ark waterproof (Hebrew: *kopher, kaphar*). Interestingly, the word used here for pitch is later translated as atonement (Leviticus 17:11). In providing a protective covering against the waters of judgment, the Ark becomes a beautiful picture of our salvation through Christ's atonement.

The Ark not only saved Noah and his family, but also preserved animal life (Genesis 6:19). Probably the most familiar aspect of the story of Noah is the animals. Noah was commanded to take two of every kind of bird, animal, and creeping thing of the ground, male and female (seven pairs of all clean animals and the birds, and a pair of the unclean animals) "to keep seed alive upon the face of all the earth" (Genesis 6:19-21; 7:1-3). God's purpose for the Ark was to "keep seed alive" on the earth, a statement meaningful only in the context of a universal flood. If the Flood was local, many people and animals would have continued to live throughout other regions of the world. But according to the testimony of Scripture, all life on earth was destroyed by the Flood, as "Noah only remained alive, and they that were with him in the ark" (Genesis 7:23). So, just as Noah and his family were the only people that remained on the earth, the only animals left on the earth were those that were on the Ark.

The Ark is a picture of salvation, as God saved only Noah and his family and preserved animal life from the global judgment that brought utter destruction to the entire world.

EVIDENCE 3
ALL SCRIPTURE TESTIFIES OF A GLOBAL FLOOD

G enesis clearly and explicitly teaches that the Flood of Noah's day was worldwide and catastrophic. But, is anything said in the rest of the Bible that would confirm that interpretation? Actually, the global nature of the Flood is validated in many places throughout both the Old and New Testaments.

The Old and New Testaments Validate the Global Flood

It may be a surprise to learn that numerous authors refer to the great Flood of Noah's day, including Job, David, Isaiah, Matthew, Luke, the author of Hebrews, and Peter. When Job addressed God's sovereign control over nature and the destructive powers of nature, he talked about the waters that God sent out that overturned the earth (Job 12:15). David also mentioned the Flood in Psalm 29:10: "The LORD sitteth upon the flood; yea the LORD sitteth King for ever." The Hebrew word used here for flood is *mabbul*. This particular word for flood is only used here and in Genesis 6–9 in reference to Noah's Flood. The earth today is drastically different from the earth that existed before the Flood (see Evidence 4). The geological changes that occurred due to the recession of the flood waters are referred to in Psalm 104:5-9.

Who laid the foundations of the earth, that it should not be removed for ever. Thou coveredst it with the deep as with a garment: the waters stood above the mountains. At thy rebuke they fled; at the voice of thy thunder they hasted away. They go up by the mountains; they go down by the valleys unto the place which thou hast founded for them. Thou hast set a bound that they may not pass over; that they turn not again to cover the earth.

Does the last verse seem familiar? Interestingly, this section of the Psalm describing the Flood ends with the promise given to Noah that the waters would not again cover the earth. Isaiah also referred to the promise made to Noah, saying "that the waters of Noah should no more go over the earth" (Isaiah 54:9). He made this statement to assure the Israelites that God's wrath, like the Flood, would subside. Because of God's love, compassion, and mercy, He would uphold His covenant.

Hebrews 11 is the popular chapter on faith, in which men and women are recognized for their living faith in God and in His commands. Noah is among those recognized.

Genesis can be trusted as accurate historical narrative.

By faith Noah, being warned of God of things not seen as yet, moved with fear, prepared an ark to the saving of his house; by the which he condemned the world, and became heir of the righteous which is by faith. (Hebrews 11:7)

Note the universal nature of the Flood's judgment as the *world* was being condemned. This one verse in the midst of the great testimonies of faith validates the historicity of Noah and the Flood. The subject of a myth or legend would not be included in this remarkable list of persons of faith. Genesis can be trusted as accurate historical narrative.

Jesus Spoke of the Global Flood

Both the gospels of Matthew and Luke validate the global Flood of Noah through the testimony of Jesus Christ. When Jesus taught about His eventual return and judgment, He pointed to a historical Noah and Flood. He compared His return and future judgment with God's judgment accomplished through the global Flood in the past.

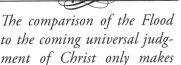

The comparison of the Flood to the coming universal judgment of Christ only makes sense if the Flood was worldwide judgment.

> But of that day and hour knoweth no man, no, not the angels of heaven, but my Father only. But as the days of Noah were, so shall also the coming of the Son of man be. For as in the days that were before the flood they were eating and drinking, marrying and giving in marriage, until the day that Noe entered into the ark, And knew not until the flood came, and took them all away; so shall also the coming of the Son of man be. (Matthew 24:36-39)

> And as it was in the days of Noe, so shall it be also in the days of the Son of man. They did eat, they drank, they married wives, they were given in marriage, until the day that Noah entered into the ark, and the flood came, and destroyed them all. (Luke 17:26-27)

Notice the global nature of Jesus' words ("took them all away," "destroyed them all"). The comparison of the Flood to the coming universal judgment of Christ only makes sense if the Flood was worldwide judgment.

Peter Writes of the Global Flood

Peter provided additional validation of the historicity of the Genesis account in the New Testament. In Peter's teaching on the coming judgment, he included numerous references to Noah's Flood.

Which sometime were disobedient, when once the long-suffering of God waited in the days of Noah, while the ark was a preparing, wherein few, that is, eight souls were saved by water. (1 Peter 3:20)

And spared not the old world, but saved Noah the eighth person, a preacher of righteousness, bringing in the flood upon the world of the ungodly. (2 Peter 2:5)

For this they willingly are ignorant of, that by the word of God the heavens were of old, and the earth standing out of the water and in the water: Whereby the world that then was, being overflowed with water, perished: But the heavens and the earth, which are now, by the same word are kept in store, reserved unto fire against the day of judgment and perdition of ungodly men. (2 Peter 3:5-7)

Interestingly, Peter used the unique Greek word *kataklusmos* to refer to the Flood. This word applies solely to Noah's Flood, indicating that this flood was different and unparalleled from other floods. This is the word from which we get our English word "cataclysm." The Genesis Flood is always referred to as the "cataclysm" in the New Testament, including in Matthew 24:39, Luke 17:27, and 2 Peter 2:5; 3:6.

The worldwide impact of the Flood is also emphasized by Peter. What do statements such as "and spared not the old world" and "the world that then was...perished" suggest? Peter affirmed that the world that existed before the Flood no longer exists. The world we live in today is vastly different from the pre-Flood world *because of* the worldwide, catastrophic Flood.

The global nature of the Flood is also emphasized in relation to the future judgment. Like the testimony of Jesus, Peter connected the future day of judgment with the judgment of the world by the Flood.

The coming judgment will encompass the entire heavens and earth, as we "look for new heavens and a new earth" (2 Peter 3:13). If the Flood was only local, only part of the earth is now "reserved unto fire" and will "melt with fervent heat" (2 Peter 3:7, 12). Peter's words are clear. Noah's Flood was global and it was cataclysmic, just as the judgment will be at the return of Jesus Christ.

If the Genesis account of the Flood is dismissed as myth or legend or downgraded to a local flood despite the clear and explicit global nature of the account, how should the other books of the Bible be treated that validate the global nature of the Flood as a historical event? The historicity of Noah is validated in Numbers, Joshua, 1 Chronicles, Isaiah, Ezekiel, Matthew, Luke, Hebrews, 1 Peter, and 2 Peter. The global nature of the Flood is affirmed by authors in the Old Testament and the New Testament, and taught by Jesus and Peter in relation to the coming judgment at the return of Jesus.

A global flood implies judgment upon all mankind. A local flood implies partial judgment. Both Jesus' and Peter's testimony show that the coming judgment will be like the past judgment—universal. The first judgment was by water, the second will be by fire. The Genesis account cannot be dismissed as myth or allegory, or be "reinterpreted" to mean a local flood. Genesis, including the global and catastrophic Flood, is the foundation of the Bible.

The Genesis account cannot be dismissed as myth or allegory or "reinterpreted" to mean a local flood.

EVIDENCE 4
GEOLOGY REVEALS THE FACT OF A GLOBAL FLOOD

The Bible clearly teaches that the Flood of Noah's day was worldwide and catastrophic. But just how destructive was it? The Bible provides evidence that the world was completely changed. But is there any scientific evidence that a global, catastrophic flood occurred? What kind of evidence would we expect to find if the global Flood actually happened?

First, we would expect to find catastrophic deposition of rocks and fossils. Second, we would expect to find deposition occurring on a regional or continental scale, not local. Geologically, this is exactly what is found. A proper interpretation of the rocks and fossils attests to a global, dynamic, watery catastrophe. The scientific evidence supports a global, catastrophic flood—Noah's Flood.

The Rocks Reveal a Global Flood

Many claim that the rock layers, the strata systems that comprise the standard geologic column, contain the major "proof" of the evolutionary theory. This is promoted as evidence for evolutionary processes and evolution's timescale (billions of years). It is claimed that each rock layer took millions of years to form. A key component for this argument of billions of years of earth history is evolutionary uniformitarianism. This is the concept that all things have developed into their present form by the same slow, natural processes that function at present, but acting over billions of years.

Are these interpretations correct? After all, they are presented as truth in classrooms around the world. In reality, though, there are glaring errors with this theory and these interpretations.

Amazingly, the rock layers presented as evolutionary evidence really only exist in textbooks. In fact, the entire geologic column, composed of complete rock layers, exists only in the diagrams drawn by geologists. Data from continents and ocean basins show that the ten strata systems of the geologic column are poorly represented on a global scale. All types of rocks (e.g., shale, limestone, granite, etc.) are found throughout the various deposits. Hundreds of locations are known where the order of the rock layers identified by geologists does not match the order of the geologic column that is taught as evolutionary fact.

The earth's crust is made of sedimentary rocks (e.g., sandstones, shales, limestones, etc.). In almost all cases, these rock layers were originally formed underwater. The layers were deposited after rapid erosion and transportation by water from various sources. But how long does it take for a rock layer to form? Evolutionists contend that it takes millions of years. However, there is no evidence that these layers took millions of years to form.

However, there is extensive evidence that these rock layers formed very quickly and under catastrophic conditions. There is no physical evidence of time boundaries between successive rock layers. If it took millions of years for them to form, it would be expected that erosion would occur at the top of one layer before the next layer was laid down on top of it. Hydraulic evidence actually shows rapid deposition of each rock layer and the continuous formation of every sequence of strata.

The rock layers supposedly represent the various "ages" of

earth history. However, many features seem to occur indiscriminately throughout the various layers. For instance, rocks in all layers exhibit similar geological formation (e.g., faults, thrusts, rifts, and folds). These distortions in the layers support their formation by a catastrophic event, not through slow and gradual processes. Additionally, common geological features of the rock layers show that "old" rocks and "young" rocks are in every formation with no physical characteristics unique to age. The scientific evidence suggests that they were all formed during essentially the same brief period of time. This is what would be expected as a result of a global, catastrophic flood.

The scope of these rock layers is also informative. The regional or continental coverage of the layers indicates a catastrophe greater than a local event. These rock layers cover immense territory. For example, the same rock layer is found in Arizona, Utah, Wyoming, Montana, Colorado, South Dakota, the Midwest, the Ozarks, and in northern New York. Equivalent formations are found across wide portions of Canada, eastern Greenland, and Scotland.

Scientific evidence indicates rock layers formed rapidly under catastrophic conditions, contradicting the notion that the layers formed over millions of years. Scientific evidence shows that rock layers share common characteristics, indicating they formed at the same time and do not represent unique ages of earth history. Continental coverage of the rock layers suggests formation by an event that was global in nature. Based on the scientific evidence, the rock layers (geologic column) seem to be the product of continuous rapid deposition of sediments. Under what conditions is all of this possible on a global scale? These data strongly suggest that the rock layers were the result of a time when "the world that then was, being overflowed with water, perished" (2 Peter 3:6).

The scientific evidence suggests that they were all formed during essentially the same brief period of time.

It is evident that the rock layers that cover continental regions formed rapidly under catastrophic conditions. The scientific evidence supports rapid, catastrophic processes—the sort of events that would have occurred during Noah's Flood, not the long ages of evolution. Although "interpreted" by evolutionary science as evidence for long ages of earth history, the scientific data in the rocks provide evidence that validates the testimony of Noah's Flood—a global, catastrophic flood.

Mountain Ranges Attest to a Global Flood

How can mountain ranges attest to a global Flood? At some point in the past, all the mountains of the world were underwater, a conclusion that is supported by science. Sedimentary rocks and marine fossils exist near their summits. These mountain ranges are full of marine fossils. Clam fossils are even found at the summit of Mount Everest!

Also, most volcanic mountains with their pillow lavas seem largely to have formed underwater. Most of the world's well-known mountain ranges today, including the Himalayan range, the Alps, the Rockies, the Appalachians, the Andes, and others, are actually composed of folded and fractured layers of ocean-bottom sediments now at high elevations. These are the kinds of deposits we would expect to find from the worldwide, world-destroying Flood of Noah's day. The scientific data indicate that all present sedimentary mountains were deposited as sediments on the ocean floor during the Flood and uplifted at the end of it.

Many contend that a global flood was not possible because there is not enough water to cover all of the vast mountain ranges. But remember, the Flood did not have to cover the present earth. It only had to cover the pre-Flood earth. The world that we know is not the same as the world prior to the Flood. It was radically altered by that global event. A pre-Flood world with lessened topographic extremes could have been covered by the Flood.

In agreement with the scientific data, Noah's Flood did not have to cover all these tall mountain ranges. Instead, Noah's Flood *caused* today's high mountains and deep ocean basins, making such a flood impossible to repeat, just as promised by God in Genesis.

Earth's Geological Features Demand a Global Flood

Do the geological features of the world today support the idea of a global, catastrophic flood? What prominent geological feature comes to mind when you think of America? Most people probably immediately think of Grand Canyon. Grand Canyon is a remarkable sight. One obvious feature of Grand Canyon is all of the rock layers (strata). About 5,000 feet of layered strata are stacked on top of one another in Grand Canyon. But how was it formed and how long did it take? Did the Colorado River, migrating back and forth for 65 million years, coupled with side canyon erosion, carve out this immense gorge? What evidence is provided by the rocks?

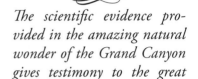

The scientific evidence provided in the amazing natural wonder of the Grand Canyon gives testimony to the great Flood of Noah's day.

The scientific evidence shows that the sedimentary rock layers were laid down underwater, with a large volume of water rushing through the area at a high velocity. Also, many of the layers show significant deformation, with some bent from horizontal to vertical in a space of 100 feet. These pronounced uplifts show that the rock layers must have been deposited over short periods of time, before the soft, muddy sediments had time to harden. So instead of pointing to millions of years of evolutionary processes, the features of Grand Canyon are best explained by a massive flood—like the one recorded in Genesis. If Grand Canyon was formed by Noah's Flood, then all the rock layers were deposited in one year!

Can the observed rock layers and the immensity of Grand Canyon really be explained by the year-long global Flood of Noah's day? Consider Burlingame Canyon in Washington. It measures 1,500 feet long, up to 120 feet deep, and 120 feet wide. Although a much smaller canyon than Grand Canyon, it was observed to form in fewer than six days. Within these few days, nearly five million cubic feet of silt, sand, and rock were removed by the cascading water. Yes, canyons can form rapidly.

In 1980, God provided a demonstration of the earth-restructuring capabilities of flowing water. The eruption of Mount St. Helens, a modern, local catastrophe, gave us a glimpse of the earth's geologic power. Most of the damage done by Mount St. Helens was water-related. Mount St. Helens had been covered by a glacier, and when it got hot, water raced down the mountain as a mighty flood, eroding soil, rocks, trees, and anything in its path. All of this was eventually deposited at the foot of the mountain. Volcanic episodes added to the fury. When the eruption was finished, up to 600 feet of sediments had been deposited, full of plant and animal remains. The sediments have hardened into sedimentary rock. A deep canyon was carved out, forming a "little Grand Canyon."

Many features that geologists are taught take long ages to form were seen to happen rapidly, including the formation of rock layers (strata), erosion of canyons, deposition of logs in upright positions, burial of peat, and glacier formation.

The scientific evidence provided in the amazing natural wonder of the Grand Canyon gives testimony to the great Flood of Noah's day. Sedimentary rocks that abound all over the world give evidence of having been formed by rapid and continuous depositional processes. Numerous other geological features observed today are best understood in relation to a global flood. There is worldwide evidence of recent water bodies in what are now desert areas. There are worldwide occurrences of raised shorelines and river terraces, and valleys are much too large for their present rivers and streams.

The earth's geological features were fashioned largely by rapid, catastrophic processes that affected the earth on a global and regional scale. The geological evidence demands a catastrophic event. Noah's Flood would have accomplished abundant geologic work. It would have resulted in the erosion and re-deposition of sediments, continents being pushed up, and plateaus being elevated. The global Flood laid down the rocks and reshaped the earth. The earth today is vastly different from the pre-Flood world, just as Peter teaches.

EVIDENCE 5
THE FOSSIL RECORD REQUIRES A GLOBAL FLOOD

The geologic column and the fossil record provide a record of the unobserved past. But what does that record reveal? Evolutionist point to it as "proof" for the long ages of evolution (billions of years). Those promoting evolutionary science contend that the fossil record shows the long ages of evolutionary development and the rock layers contain the life forms that were in the different "ages." In this scenario, the formation of the rocks and fossils is the result of either slow and gradual processes or occasional rapid processes of local extent.

Those that contend that the Flood of Noah's day was global expect the fossil record to show evidence of catastrophic deposits of rocks and fossils on a regional or continental scale. But, which position does the scientific data actually support? When the fossil record is studied, does it provide evidence for long ages of evolutionary development, or evidence for a catastrophic watery event in the recent past? These obviously are two mutually exclusive positions.

The Fossil Record Shows Global Watery Burial

Do the places where fossils are found and the ways they develop give evidence of the events that caused their formation? Fossils are typically found buried in sedimentary rocks, which are formed when sediments are deposited by moving water and then harden. As they harden into sedimentary rock, the dead things within them harden into fossils. The study of rock layers gives evidence of catastrophic water-deposited layers that cover huge regions and some continents.

Does the fossil record support evolutionary science's claims, or a worldwide flood? Actually, the composition of the fossil record is very telling. The actual fossil record, from top to bottom, in every layer, all over the earth, overwhelmingly consists of shallow-water marine fossils. As it turns out, 95 percent of all fossils are shallow marine invertebrates, mostly shellfish. For instance, clams are found in the bottom layer, the top layer, and every layer in between. There are many different varieties of clams, but they are in every layer, and many of those varieties are still alive today. The same could be said for corals, jellyfish, and many others. The fossil record documents primarily *marine* organisms buried in *marine* sediments that were catastrophically deposited. How did so many marine fossils end up on the land? Which claim does this evidence support?

Of the other 5 percent of the fossils, by far most are plants. Land-dwelling animals, such as mammals and dinosaurs, are poorly represented in the fossil record. All of the vertebrate fossils considered together (fish, amphibians, reptiles, birds, and mammals) comprise only 0.0125 percent of the entire fossil record, and only 1 percent of these (or 0.000125 percent of the total) consist of more than a single bone! Surely, the vertebrate fossil record is far from complete. However, land vertebrates are what are often pictured in evolutionary fossil charts in textbooks. The case for evolution is made from them, yet they do not accurately portray the real fossil record.

The fossil record supports a global catastrophic event, as fossils are found distributed over the entire surface of the earth. The deposited marine fossils, including bottom-dwelling marine life forms, are all predominantly found on today's continents, in every conceivable ng broken, mangled, disarticulated, and lumped together, as well as separated by size and type.

Interestingly, the fossil record shows animals from mixed habitats grouped together. Vast graveyards of fossil deposits can be found all over the world. These fossil graveyards often contain creatures of every kind and from every environment. Marine life, dinosaurs, lizards, fish, mammals, and plant life are all found buried together. These fossil graveyards suggest cataclysm, the rapid covering of the land with ocean waters that violently killed, washed together, buried, and rapidly fossilized animals from various ecosystems.

Dinosaur Soft Tissue Supports a Recent Cataclysmic Event

One of the discrepant positions between evolutionary science and the global Flood of Noah's day is the question of when the fossils formed. According to biblical records, Noah's Flood would have laid down the rocks and fossils only about 4,500 years ago. This obviously is in stark contrast with the evolutionary timescale, which claims that the geologic column and fossils are a record of earth's millions-of-years prehistoric past. Most notably, evolutionists point to the Mesozoic era (Triassic, Jurassic, and Cretaceous periods) as defining the "age of the dinosaur." Does the fossil evidence substantiate these claims, or are data being misinterpreted or ignored to preserve evolutionary presuppositions?

Over the last few years, evidence supporting a *recent* rapid and cataclysmic burial has been discovered that has left evolutionists scrambling for an explanation (or inventing nonsensical explanations)—namely, the discovery of fossils containing soft tissue, including soft tissue from dinosaurs. To date, over 40 examples of original soft tissue have been discovered and reported within the scientific community.

This soft tissue includes original skin, connective tissue, blood vessels, and blood cells. The highest profile discovery of soft tissue was the finding of blood vessels and blood cells in a *T. rex* femur that was supposedly 70 million years old. This puts evolutionists in a huge quandary, since scientific experiments reveal that organic materials such as these should not be preserved beyond 100,000 years (maximum). In fact, collagen, such as blood vessels, cannot survive longer than 10,000 years, even in optimal conditions. The scientist who made the discovery, an evolutionist, even contends that "translucent blood vessels...by all the rules of paleontology should have long since drained from the bones. It's a matter of faith among scientists that soft tissue can survive at most for a few tens of thousands of years, not the 65 million since *T. Rex* walked what is now the Hell Creek Formation in Montana."[2]

The scientific data of the fossil record—water burial, rapid formation, catastrophic event/processes—are best explained by a recent, global, and catastrophic flood, like the Flood of Noah's day. Dr. Henry Morris III summarizes the importance of a biblical understanding of the Flood as it relates to geology, including the geologic column and the fossil record:

Both the Bible and science, rightly interpreted, point to a global, catastrophic flood.

> The creation model thus also includes as another major component the global cataclysm of the Deluge. It maintains that the actual facts of geology, including the sedimentary rocks and their fossils, as well as the present structure of the earth's crust and surface features, can be more easily and naturally explained in terms of the Flood than they can in terms of the uniformitarian model. The various rock systems do not represent evolutionary ages at all, but rather diluvial stages.[3]

It is not a matter of science versus the Bible. Both the Bible and science, rightly interpreted, point to a global, catastrophic flood as the true explanation for the sedimentary fossil-containing strata. The various phenomena related to the Flood (hydraulic activity, volcanism,

---

2. Yeoman, Brian. Schweitzer's Dangerous Discovery. *Discover Magazine,* April 27, 2006.
3. Morris, Henry M. and Henry M. Morris III. 1996. *Many Infallible Proofs.* Green Forest, AR: Master Books, 288-289.

tectonic upheavals, and subsequent winds and glaciers) provide the most effective model to correlate with the evidence in the geologic strata.

WHY A GLOBAL
FLOOD MATTERS

A flood of the proportions described in Genesis would have resulted in vast amounts or erosion and re-depositing of sediments, fossilization of plants and animals, volcanism, and re-distribution of radioisotopes. If the Flood happened the way the Bible says it happened, then it laid down the rocks and fossils, and there is no remaining "evidence" for an old earth—or evolution, for that matter. Using the rocks and fossils to support long ages of evolutionary time requires denial of the biblical teaching of the Flood of Noah's day and a misinterpretation of what is truly revealed scientifically in the rocks and fossils. A proper interpretation of the scientific evidence found in the rocks and fossils points to a worldwide, dynamic, watery catastrophe: the global, catastrophic Flood.

But does it really matter whether we hold to a global or local flood of Noah's day? Is this just an insignificant theological debate? Is this only relevant to those really interested in science? Evolutionists, of course, consider the Flood account in Genesis to be a myth. However, recently many in the Christian community have also begun dismissing the reality of the Flood event, or downgrading the extent of Noah's Flood to a local event. These compromise positions are the result of Christians placing "man's science" over the authority of the Bible, attempting to incorporate evolutionary science into the book of Genesis. The "reinterpretation" of Genesis as portraying a local flood is part of a greater attack on the validity, reliability, and authority of Genesis, all of which result in significant theological compromises to God's Word, God's nature, and the Gospel message.

The extent of Noah's Flood is not merely an insignificant theological debate, but rather a significant doctrine of Christian theology. In addition, accepting or rejecting the global nature of Noah's Flood has immense ramifications on numerous scientific disciplines, especially geology and paleontology. In the seminal book that launched the creation science movement, John Whitcomb and Henry Morris attested to the importance of the doctrine of a global flood, both scientifically and theologically.

> The question of the historicity and the character of the Genesis Flood is no mere academic issue of interest to a small handful of scientists and theologians. If a world-wide flood actually destroyed the entire antediluvian human population, as well as all land animals, except those preserved in a special Ark constructed by Noah (as a plain reading of the Biblical record would lead one to believe), then its historical and scientific implications are tremendous. The great Deluge and the events associated with it necessarily become profoundly important to the proper understanding of anthropology, of geology, and of all other sciences which deal with historical and prehistorical events and phenomena.

> But of even greater importance are the implications of the mighty Flood of Genesis for Christian theology. For the universal catastrophe speaks plainly and eloquently concerning the sovereignty of God in the affairs of men and in the processes of nature. Furthermore, it warns prophetically of a judgment yet to come, when the sovereign God shall again intervene in terrestrial events, putting down all human sin and rebellion and bringing to final fruition His age-long plan of creation and redemption.[4]

---

4.   Whitcomb, John C. and Henry M. Morris. 1961. *The Genesis Flood*. Phillipsburg, NJ: Presbyterian and Reformed Publishing, xxxv.

It is not merely an insignificant theological or scientific debate. Both the biblical and scientific evidence clearly attest to a global Flood. And the scientific and theological implications are critical to a proper understanding of the world we live in and of our Christian faith. Do not compromise with evolutionary science, which is rooted in rejecting God, denying His Word, and dismissing the Gospel message. Commit to upholding the inerrancy, infallibility, and the authoritative truth of God's Word in its entirety, including Genesis—the foundation of the Bible and of the Gospel message.

> If I profess with the loudest voice and clearest exposition every portion of the truth of God except precisely that little point which the world and the devil are at that moment attacking, I am not confessing Christ, however boldly I may be professing Christ. Where the battle rages, there the loyalty of the soldier is proved, and to be steady on all the battlefield besides, is mere flight and disgrace if he flinches at that point.
>
> — Martin Luther

ABOUT THE AUTHOR

 Dr. Brad Forlow received his B.S. in Chemical Engineering at Florida Institute of Technology, and his Ph.D. in Chemical Engineering at the University of Oklahoma. For four years he held the post of Assistant Professor of Research at the University of Virginia before working in pharmaceutical research for an additional six years, most recently for Wyeth/Pfizer. In addition to his science training, Dr. Forlow is completing his theological training at Southwestern Baptist Theological Seminary. Dr. Forlow currently serves on the life sciences research team at the Institute for Creation Research in Dallas, Texas, as well as functioning as Associate Science Editor at the Institute. He is married to Dr. Rhonda Forlow, who serves as ICR's K-12 Education Specialist. The Forlows have three children and reside in Dallas.

FOR MORE INFORMATION

Sign up for ICR's FREE publications!

Our monthly *Acts & Facts* magazine offers fascinating articles and current information on creation, evolution, and more. Our quarterly *Days of Praise* booklet provides daily devotionals—real biblical "meat"—to strengthen and encourage the Christian witness.

To subscribe, call 800.337.0375 or mail your address information to the address below. Or sign up online at www.icr.org.

Visit ICR online

ICR.org offers a wealth of resources and information on scientific creationism and biblical worldview issues.

- ✓ Read our daily news postings on today's hottest science topics
- ✓ Explore the Evidence for Creation
- ✓ Investigate our graduate and professional education programs
- ✓ Dive into our archive of 40 years of scientific articles
- ✓ Listen to current and past radio programs
- ✓ Order creation science materials online
- ✓ And more!

*Visit our Online Store at www.icr.org/store
for more great resources.*

INSTITUTE
for CREATION
RESEARCH

P. O. Box 59029
Dallas, TX 75229
800.337.0375